Guide to the Plants of Granite Outcrops

Guide to the Plants of Granite Outcrops

William H. Murdy

M. Eloise Brown Carter

University of Georgia Press

Athens and London

Designed by Sandra Strother Hudson
Set in Galliard
Printed and bound by Sung In
Printing, America Inc.

The paper in this book meets the
guidelines for permanence and
durability of the Committee on
Production Guidelines for Book
Longevity of the Council on Library
Resources.

Printed in South Korea

04 03 02 01 00 C 5 4 3 2 1

Library of Congress
Cataloging-in-Publication Data
Murdy, William H., 1928–
Guide to the plants of granite
outcrops / William H. Murdy,
M. Eloise Brown Carter.
p. cm.
Includes bibliographical references
and index.
ISBN 0-8203-2133-8 (alk. paper)
1. Botany—Georgia. 2. Botany—
Piedmont (U.S. : Region)
3. Phytogeography—Georgia.
4. Phytogeography—Piedmont
(U.S. : Region) 5. Granite outcrops—Georgia. 6. Granite outcrops—Piedmont U.S. : Region)
7. Plants Identification. I. Carter, M.
Eloise Brown, 1950– . II. Title.
QK155.M87 2000
581.9758—dc21 99-14453

British Library
Cataloging-in-Publication Data
available

To Nancy

W. H. M.

To Stefanie and Cindy

M. E. B. C.

Contents

Acknowledgments

We wish to thank James N. Skeen, for his major contribution to the photography; Vicky Holifield, for her encouragement and the illustrated glossary; the late Bob Platt, who was the first to propose an illustrated guide to granite outcrop plants; Martha Jane Murdy for her editorial help; Sheilah Conner, for her patience in working on the manuscript; and Emory University for financial support of the project. We are also grateful to the following persons for the use of plant photographs: Eloise Carter (*Amelanchier arborea, Erythronium americanum, Saxifraga virginiensis, Amsonia tabernaemontana, Prunus umbellata, Tradescantia hirsuticaulis, Agave virginica, Coreopsis grandiflora, Lechea racemulosa, Linum virginianum, Rhynchospora grayi, Talinum teretifolium, Agalinus tenuifolia, Panicum lithophilum*); Wilbur Duncan (*Krigia virginica*); Vicky Holifield (*Lepuropetalon spathulatum*); William Murdy (*Parthenocissus quinquefolia, Belamcandra chinensis, Hypericum prolificum, Talinum mengesii*); Robert Platt (*Amphianthus pusillus, Portulaca smallii*); and Harvey Young (*Celtis tenuifolia*). All other photographs were provided by James Skeen.

Introduction

Three rock mounts, Stone Mountain, Mount Arabia, and Panola Mountain, are striking features of the rolling landscape east of Atlanta. These massive domes are unusual in a region where flat or gently sloping exposures, called "flatrocks," are more common. Such rock exposures occur both in clusters and distantly removed from one another. They range in size from several square meters to hundreds of acres. From a bird's-eye view, rock outcrops appear as islands in a landscape of forest, fields, and development.

The Piedmont Region of the southeastern United States is underlaid with crystalline rock of Precambrian age, most of which is covered with a soil mantle of variable depth but where occasionally, at intervals from Virginia through North and South Carolina into Georgia and Alabama, bedrock is exposed at the surface. These rock exposures, depending on their composition and degree of metamorphosis, may be classified as biotite gneiss, mica-schist, or granite but are referred to herein as granite. Over 90 percent of the estimated twelve thousand acres of outcrops in the Piedmont Region are located in Georgia.

Unique assemblages of plants and animals that have adapted to these extreme environments are associated with the rock exposures. The surfaces of the rock outcrops are not bare but are instead darkened by an array of microscopic lichens. Two conspicuous plants that grow directly on rock surfaces are *Parmelia conspersa*, a crustose lichen, and *Grimmia laevigata*, the Rockmoss. During periods of dryness, the Rockmoss appears to be dead, but within hours of a rain these plants become emerald green and functional. These mosses catch soil particles and over time a mat is built up under the plants. Deeper soil enables the establishment of larger lichens belonging to the genus *Cladonia,* such as Reindeer Moss (*C. rangiferina*) and the hairy cap moss

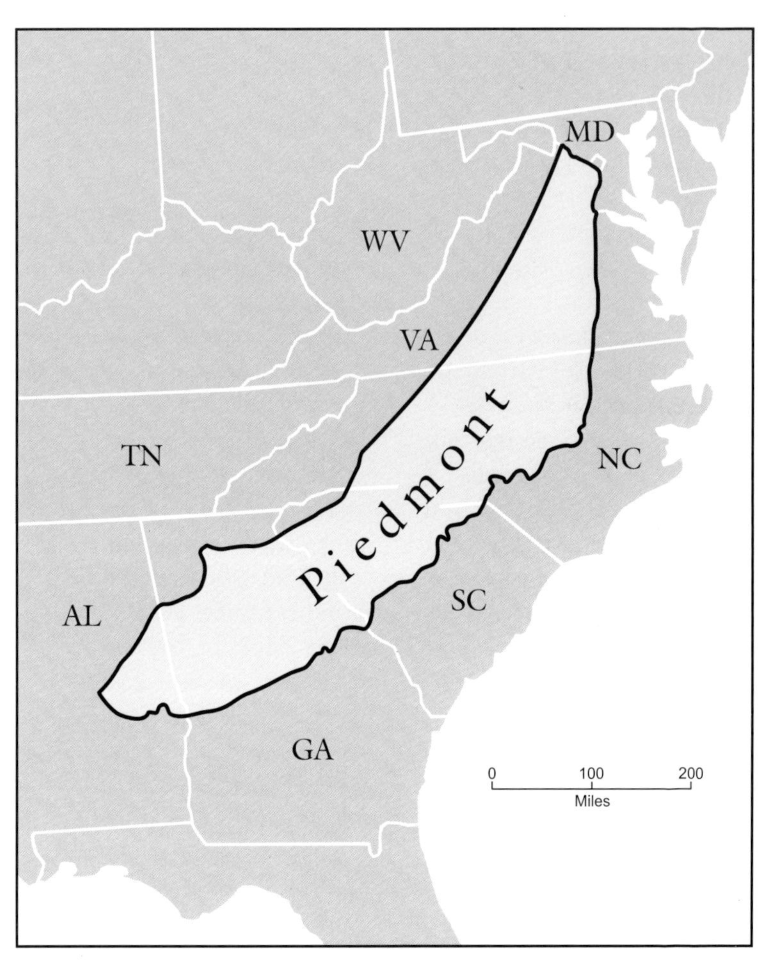

The Piedmont Region of the southeastern United States, where granite outcrop plants occur

(*Polytrichum commune*). These lichens and mosses are later followed by a variety of vascular plants.

Granite outcrops afford a number of unusual habitats for plants and support a unique flora, including species found nowhere else in the world. Habitats available to higher plants include depressions in the rock containing variable amounts of soil, vernal pools, rock crevices, rubble heaps resulting from past quarrying, and border areas between the rock and adjacent forest or field.

Although the longevity of any single exposure may not be great, outcrops have been open to plant habitation for thousands of years. For example, the rare Pool Sprite (*Amphianthus pusillus*) has had time to become adapted to an exceedingly specialized and infrequent habitat comprised of vernal pools of a critical size, topography, soil depth, and water retention capacity. Furthermore, it has diverged from close relatives to the extent that it is now the only species in a genus of uncertain taxonomic affinities.

While granite outcrops occur throughout most of the Piedmont, the greatest concentration of exposed rock is east of Atlanta in DeKalb, Rockdale, and Walton counties. Local landowners and governments generally do not recognize or appreciate these unique plant and animal ecosystems that are specially adapted to these desertlike, isolated environments. In fact, throughout much of the region outcrops are considered to be wastelands, unfit for grazing or construction. Consequently, they often become dumping grounds for refuse and discarded household items. Boats, mattresses, furniture, dead livestock, clothing, and construction waste can be found on disturbed outcrops that are accessible by roads.

Protection of some of the most undisturbed sites has been afforded primarily by inaccessibility or by the conscientious efforts of enlightened landowners. What is most troubling is the increasing damage to outcrops as suburban development spills over into rural areas. Outcrops become loading areas for logging and pathways for moving and parking heavy equipment. The resulting destruction could be prevented by advanced planning and protection of outcrops during development. By their own unique and varied beauty, these natural areas could enhance the character of residential and commercial developments. Efforts to educate the public and to protect these vulnerable ecosystems must continue.

Preservation of some of the larger, least disturbed granite outcrops is already in progress. Ten outcrops in Georgia currently have protection status: Stone Mountain State Park in DeKalb County, just east of the town of

The Confederate Daisy flowers abundantly in the fall at Stone Mountain Park, east of Atlanta.

Stone Mountain; Panola Mountain State Conservation Park in Rockdale County, southeast of Atlanta; Camp Meeting Rock in Heard County, between Franklin and the Alabama state line; Davidson-Arabia Mountain Park, which belongs to DeKalb County and is situated at the southeast corner of the county; Heggie's Rock, supervised by The Nature Conservancy and located in Columbia County; the Newton County Reservoir, north of Covington; the Rockdale County Big Haynes Creek Nature Center, north of Conyers; Little Kennesaw Mountain in Kennesaw Mountain National Battlefield Park in Cobb County; the Charlie Elliott Wildlife Center in Jasper County; the State Arboretum of Georgia in Jackson County; and Tribble Mill Park in Gwinnett County.

In South Carolina, the best known granite outcrop, called Forty Acre Rock in Lancaster County, is owned by the South Carolina Department of Natural Resources and is therefore protected. Granite outcrops in North Carolina with protected status include: Mitchell's Mill, Wake County, part of a state park; The Rock, located in Wake County and owned by the Triangle Lands Conservancy; Salem Lake Natural Area, Forsyth

Spectacular displays of spring wildflowers are arrayed in rings of color in depression pits on the rock surface.

County, which is owned by the city of Winston-Salem and has several small outcrops; and Rocky Face in Alexander County, owned by the National Forest Service.

Visits to granite outcrops during different seasons of the year reveal strikingly varied vistas. In the winter, when conditions are cool and moist, mosses and lichens are conspicuous. Several spring flowering annuals still in the vegetative state, such as the red succulent *Diamorpha*, are abundant in the shallow soil islands and around the rock borders. In some of the rock-rimmed pools one may observe Pool Sprite and Black-Spored Quillwort, both rare and endemic to the granite outcrops.

The peak flowering season is spring, when a number of shallow-rooted winter annuals come into flower along with deep-rooted perennials. Spring is also the best time to observe the striking zonation of species in the island communities. The distribution of plants forms rings of color in which species occupy different zones based primarily on soil depth, which in turn determines the amount of moisture available for plants. In 1964 Burbanck and Platt showed that the shallowest soils, averaging 5.5 cm., are occupied by the Diamorpha (*Diamorpha smallii*), a diminutive succulent with reddish leaves and stems, plus a few lichens and mosses. The white-flowered Sandwort (*Arenaria uniflora*), along with lichens and mosses, dominates in soils between 7 and 15 cm. deep. Bentgrass (*Agrostis elliottiana*) and the blue-flowered Toadflax (*Linaria canadensis*) may be found here also. In soils that range from 15 to 40 cm., perennial herbs such as Cottony Groundsel (*Senecio tomentosus*) and Sunnybell (*Schoenolirion croceum*) dominate.

In summer and early fall, shallow soils may be occupied by the annuals Granite Sedge (*Cyperus granitophilus*) and Small's Portulaca (*Portulaca smallii*) and the succulent perennials Prickly Pear (*Opuntia compressa*) and Rock Pink (*Talinum teretifolium* or *T. mengesii*). The deeper soils support a variety of species including the Confederate Daisy (*Viguiera porteri*) and Blazing Star (*Liatris microcephala*) that put on a spectacular show when they flower late in the season.

This guide illustrates and describes many species common to granite outcrop communities, including some rare species. Flowering plant species are arranged according to four categories of seasonal flowering: early spring, spring, summer, and late summer and fall. However, many species overlap these seasonal categories so the reader should consult adjacent sections as well for further reference. Flowering plants are followed by cone bearing plants and seedless vascular plants. Eighty species are illustrated and twenty-eight additional species are included in the descriptive narratives. Lichens and mosses are not included in the guide, nor does this work contain an exhaustive study of every species that grows on granite outcrops. Rather, this book is written for the layperson who wishes to explore rock outcrops and for anyone who is interested in native plants.

Early Spring Flowering Plants

False Garlic

Allium bivalve (L.) Kruntz
Lily Family

Description: A bulbous perennial, the False Garlic is among the most common species of the granite outcrops. Basal linear leaves are green when the plant is in flower. An umbel of three to ten flowers terminates a naked stalk that can be anywhere between 6 and 12 inches in height. Its relatively large, lilylike flowers are white or greenish-white with petals about ½ inch long.

Occurrence: In large populations where the soil of depression pits is deep enough for perennials but too shallow for shrubs. It flowers in both spring and fall.

Distribution: Southeastern United States and west to Nebraska, Texas, and Mexico

Downy Serviceberry

Amelanchier arborea (Michx.f.) Fern.
Rose Family

Description: The Downy Serviceberry is a small tree with alternate, simple, serrate leaves that are hairy beneath. Its showy white open flower clusters appear in early spring. Each flower has five petals, many stamens, and a pistil with five stigmas. Pollination produces fleshy, red to purple applelike fruits with five to eight seeds in each.

Occurrence: In deep soils that border granite outcrops. It is among the earliest trees to flower in the spring.

Distribution: Eastern North America

Pool Sprite

Amphianthus pusillus Torr.
Figwort Family

Description: The Pool Sprite is a rare species and is listed on both state and federal registers as threatened. It is wholly confined to granite outcrops. The Pool Sprite is an aquatic annual that completes its life cycle in shallow outcrop pools in winter and spring. It consists of a submerged rosette of small lance-shaped leaves and paired floating leaves borne on long, delicate stems. As the generic name infers, it bears two kinds (*amphi*) of flowers (*anthus*). Tiny white solitary flowers bloom between the paired leaves above the water, and self-fertilizing flowers that never open are borne by the submerged rosette.

Occurrence: In shallow soil in pools that hold water for several months of the year. The population exists as dormant seeds most of the year, from the time the pools dry up in mid to late spring and until they reform in winter.

Distribution: Georgia, Alabama, and South Carolina

Thimbleflower

Anemone caroliniana Walt.
Buttercup Family

Description: The Thimbleflower is a perennial that grows from a single round underground stem. Its hairy stem, which can reach up to 12 inches in height, bears a single flower and one set of opposite leaves that are dissected into linear segments. Basal leaves have three lobed segments with scalloped margins. The flowers, about 1 inch across, bear a variable number of white to bluish petal-like sepals and many stamens. At maturity, the cylindrical cluster of small dry fruits resembles a thimble, hence the name Thimbleflower.

Occurrence: Rare on granite outcrops, it typically grows in the thin, organic soil under Red Cedar trees.

Distribution: Southeastern United States from North Carolina to Mississippi

Granite Whitlow-Wort

Draba aprica Beadle
Mustard Family

Description: One of many winter annuals that grow on the granite outcrops, the Whitlow-Wort germinates in the fall and exists through winter in the form of a rosette of leaves. It develops a flowering stem between 5 and 13 inches in height in the early spring. The leafy stems bear few short branches and clusters of tiny flowers. Its small dry elliptical seed pods are densely covered with star-shaped hairs.

Occurrence: Rare on granite outcrops where it may be found growing in the partial shade of Red Cedar trees.

Distribution: Southeastern United States, where it is rare

Trout Lily

Erythronium americanum Ker.
Lily Family

Description: This bulbous perennial herb has two green leaves that are mottled with brown or purple, resembling the scale coloring of trout, hence its common name. In early spring, large nodding lilylike flowers are borne by older plants on a leafless stem that can grow up to 6 inches in height. The flowers of the Trout Lily have six separate yellow sepals and petals, and six stamens all arising from the base of a three-celled pistil.

Occurrence: Abundant in depression pits with deep soil and in moist woodland adjacent to the rock outcrop.

Distribution: Eastern parts of the United States and Canada

Yellow Jessamine

Gelsemium sempervirens (L.) Ait. f.
Logania Family

Description: This twisting vine with showy yellow, fragrant flowers becomes rooted in cracks and crevices in rock and trails along the ground, and sometimes entwines itself in trees and shrubs. The wiry stem bears opposite evergreen leaves. The large funnel-shaped flowers consist of a five-lobed corolla, five stamens attached to the inside of the corolla tube, and a pistil. Flowers may be of two forms: short styles with stigmas below the stamens and long styles with stigmas above the stamens.

Occurrence: Outcrop rock crevices.

Distribution: Eastern United States

Small Bluet

Houstonia pusilla Schoepf
Madder Family

Description: A tiny annual, the Small Bluet is composed of slender stems, 1 to 5 inches tall, emerging from a basal rosette of leaves and bearing a single flower. Each flower has a four-lobed purple (rarely white) corolla with a reddish center. The plants often grow in large populations.

A related species, *H. caerulea* (L.) Torr and Gray, Quaker Ladies, is sometimes encountered in the deeper soils of the outcrops. Quaker-Ladies is a perennial with slightly larger flowers that are light blue to white with a yellow center.

Occurrence: One of several native weedy species that are common in the shallow soils of granite outcrops.

Distribution: Southeastern United States

Dwarf Dandelion

Krigia virginica (L.) Willd.
Aster Family

Description: The Dwarf Dandelion, which bleeds with milky latex when damaged, produces a small head of dandelionlike yellow flowers that are borne on slender leafless stems arising from a basal rosette of hairy leaves. The mature nutlet fruits are topped by conspicuous scales and bristles.

Occurrence: In shallow soils of most granite outcrops.

Distribution: A native weed species widespread throughout the United States

Little People

Lepuropetalon spathulatum (Muhl.) Ell.
Saxifrage Family

Description: The Little People is an easily overlooked species due to its diminutive size. An annual, it grows in the form of a rosette that stands about 1 inch in height and breadth. Its diminutive leaves are spatulate in shape, and its tiny flowers have five white petals.

Occurrence: In moist rock or sand substrate.

Distribution: Southeastern United States from North Carolina to Mississippi

Toadflax

Linaria canadensis (L.) Dumont
Figwort Family

Description: The Toadflax is a winter annual or biennial that first develops a rosette of spreading stems with opposite linear leaves, then grows an erect stem between 8 and 30 inches in height. Its stem bears alternate linear leaves and a terminal many-flowered inflorescence whose blue snapdragonlike flowers have a long narrow spur.

Occurrence: A native weed common to roadsides, the Toadflax occurs on most granite outcrops.

Distribution: North America

Early Saxifrage

Saxifraga virginiensis Michx.
Saxifrage Family

Description: The Early Saxifrage is a perennial with ovate, shallowly serrate leaves that form a basal rosette. Its soft hairy leafless stem, between 4 to 18 inches in height, terminates in a branched inflorescence. The small flowers have five white petals and ten stamens. *Saxifraga michauxii* Britton, Mountain Saxifrage, has coarsely toothed basal leaves and flower petals unequal in size and shape. It occurs in the western outcrops of North Carolina.

Occurrence: In moist soil and partial shade in outcrop border areas where seepage occurs.

Distribution: Eastern North America

Granite Stonecrop

Sedum pusillum Michx.
Orpine Family

Description: This rare succulent annual resembles the ubiquitous *Diamorpha smallii* both in size and in the structure of its leaves, flowers, and fruits. It differs in that it has green stems and leaves and pistils that are separate to the base. While superficially similar, these two species have very different chromosome numbers. Experimental attempts to cross the plants have been unsuccessful.

Occurrence: In shallow depression pits under the partial shade of Red Cedar trees.

Distribution: Wholly confined to granite outcrop communities in Georgia and a few locations in South Carolina and North Carolina

Winged Elm

Ulmus alata Michx.
Elm Family

Description: A medium-sized tree with conspicuous corky wings on the branches, the Winged Elm has alternate leaves that are doubly serrate, hairy on the underside, and up to 3 inches long and 2 inches in width. Its flowers, which lack petals, develop in early spring and occur in short pendulous branches before the leaves appear. In only a few weeks' time, the hairy notched elliptical fruits, which can be as much as ½ inch in length, fall from the tree.

Occurrence: In depression pits having deep soil and in border areas on the granite outcrops.

Distribution: Southeastern United States

Atamasco Lily

Zephyranthes atamasco (L.) Herbert
Amaryllis Family

Description: A bulbous perennial with basal linear leaves that grow up to 16 inches in length, the Atamasco Lily produces a single erect white to pinkish flower that is borne on a leafless stem that may be as tall as 12 inches. Its petals and sepals are united into a funnel-shaped flower, 2 to 3 inches in length, having six stamens and a pistil with three stigmas.

Occurrence: In abundance in the moist soils of some granite outcrops.

Distribution: Common, especially in the Coastal Plain, from Virginia to Mississippi

Spring Flowering Plants

Bentgrass

Agrostis elliottiana Shultes
Grass Family

Description: The Bentgrass is an annual with delicate stems that terminate in an open and branched inflorescence bearing inconspicuous flowers. A slender appendage, or awn, is usually present on one bract of each spikelet enclosing the flower. The leaves are narrow, flat, and mostly basal.

Occurrence: Usually abundant in shallow depression pits of most granite outcrops.

Distribution: Southeastern United States

Wild Onion

Allium cuthbertii Small
Lily Family

Description: The stem of the Wild Onion, which can be anywhere between 12 and 30 inches in length, terminates in a many-flowered umbel whose small white blooms have petals ¼ to ⅓ inch long. The leaves are withered or absent when the plant is in flower. When bruised, the plant emits a strong onionlike odor.

A related species, *A. speculae* Owenby and Aase, Flatrock Onion, occurs rarely on outcrops in Georgia. Its leaves are narrow and its flower petals have a pinkish hue.

Occurrence: Abundant on some but not all major outcrops.

Distribution: Southeastern United States

Blue Star

Amsonia tabernaemontana Walt.
Dogbane Family

Description: The Blue Star is a perennial having alternate ovate to lanceolate leaves. Milky latex bleeds from the leaf or stem if cut. Clusters of unbranched stems, 2 to 3 feet in height, terminate in branched inflorescences bearing pale blue flowers. The five petals of each flower are fused below into a tube and flair out in pin-wheel fashion.

Occurrence: In the border area between outcrop and deciduous forest.

Distribution: Southeastern United States

Rock Cress

Arabis laevigata (Muhl. ex Willd.)
Mustard Family

Description: A biennial, the Rock Cress spends its first year in the form of a leafy rosette. In the second year of growth, a single stem develops with leafy branches above terminating with inflorescences. The leaves are usually toothed or serrate, somewhat clasping at the stem, and vary from 3 to 5 inches in length. Its flowers are small with four creamy white petals. The long narrow seed pods are curved and may reach 4 inches in length.

Occurrence: A rare species on the granite outcrops, it grows in thin, partially shaded soil.

Distribution: Eastern United States

Sandwort

Arenaria uniflora (Walter) Muhl.
Pink Family

Description: A delicate white-flowered annual, the Sandwort has a short stem that arises from a rosette of leaves that wither as flowers and fruits develop. It has small, narrow opposite leaves and bears flowers on the ends of branches. The flowers have five separate petals each with an apical indentation, ten stamens, and a pistil with three stigmas.

Two additional *Arenaria* species occur on granite outcrops. *A. alabamensis* (McCor., Bozem. and Spongb.) Wyatt, reported from outcrops in Alabama and North Carolina, is found in wetter sites and has smaller flowers than *A. uniflora*. The other species, *A. glabra* Michx., which differs from the two previous species in having larger sepals and larger pleated petals, is found in rocky locations.

Occurrence: The Sandwort is restricted to granite and sandstone outcrops.

Distribution: North Carolina to Alabama

Crossvine

Bignonia capreolata L.
Bignonia Family

Description: A woody vine with opposite compound leaves of two leaflets and a branched tendril, the Crossvine bears handsome flowers. The tubular two-lipped corolla, 2 inches long, is red to orange on the outside and yellow or red within. The pith of the stem is shaped like a cross, hence its common name.

The Trumpet Vine (*Campsis radicans* [L.] Seemann) is a related species that climbs trees by its aerial roots. Its opposite leaves are pinnately compound with many toothed leaflets. Its flowers resemble those of the Crossvine but are longer and uniformly red to orange colored. It is widely distributed in the Southeast and has the characteristics of a native weed; for example, it is quite at home in disturbed urban settings.

Occurrence: In thickets and woodlands bordering outcrops.

Distribution: Southeastern United States

Georgia Hackberry

Celtis tenuifolia Nutt.
Elm Family

Description: The Georgia Hackberry is a shrub or small tree with alternate ovate leaves between 2 and 3 inches in length and ½ inch wide. The leaf blades have three prominent veins arising from the base, and the margin of a single leaf may be in part both serrate and entire. Its inconspicuous flowers develop at the same time as the leaves. The fruits of the Georgia Hackberry are spherical in shape and orange to brown in color. They consist of a thin fleshy part surrounding a stony pit containing a single seed.

Occurrence: In border areas between rock and woodland.

Distribution: Southeastern United States

Fringe Tree

Chionanthus virginicus L.
Olive Family

Description: A spectacular sight when its masses of drooping flowers bloom in the spring, the Fringe Tree appears as a small tree or tall shrub with simple, entire, opposite leaves. Its woody twigs have raised warty lenticels, and the leaf stalks are frequently purple. Its white fragrant flowers have four long linear petals that are fused at the base.

Occurrence: An infrequent member of the granite outcrops, the Fringe Tree grows in partial shade in border areas.

Distribution: Southeastern United States

Downy Oat Grass

Danthonia sericea Nutt.
Grass Family

Description: The Downy Oat Grass is a large perennial grass with many branches arising from the base. The sheath of the leaf surrounding the stem is densely hairy. Leaf blades can grow up to 10 inches in length. Its stems, which may reach 3 feet in height, terminate in narrowly branched inflorescences bearing relatively large spikelets. Similar to oats, the lower two bracts of the spikelet exceed the rest of the spikelet in length. A close relative, *Danthonia spicata* (L.) Beauvois ex R. and S., also occurs on the outcrops and can be distinguished from the Downy Oat Grass by the absence of hairs on the leaf sheaths.

Occurrence: In gladelike areas between rock and surrounding woods.

Distribution: Widely distributed from New Jersey to Louisiana

Diamorpha

Diamorpha smallii Britt.
Orpine Family

Description: The Diamorpha is a diminutive succulent annual, 1 to 3 inches in height, with small fleshy red alternate leaves. Its flowers have four white petals and eight stamens. These plants germinate in late fall, display a bright red color through the winter months, and reach seed maturity in the spring. The ripened seeds remain in dry fruits held up by the dried stem until their release in the fall. The species is closely related to the Sedums, and some botanists prefer the name *Sedum smallii* (Britton) Ahles.

Occurrence: In abundance in the shallowest soils of granite outcrops in the Piedmont and on sandstone exposures in the Ridge and Valley areas of northwest Georgia and the Coastal Plain of Georgia.

Distribution: Virginia to Alabama and Tennessee

Swamp Privet

Forestiera ligustrina (Michx.) Poir.
Olive Family

Description: This shrub has opposite, finely serrate, simple leaves that are up to 2 inches long and have a distinct petiole. Inconspicuous flowers, lacking a corolla, are borne in axillary clusters on twigs of the previous season. The bluish-black fruits of the Swamp Privet are larger and more ellipsoidal than those of the Chinese Privet (*Ligustrum sinense*).

This species may be mistaken for Chinese Privet. However, the Chinese Privet has leaves with very short petioles and flowers with white corollas borne in inflorescences that terminate the branches.

Occurrence: Common in the shallow soils surrounding outcrops.

Distribution: Rocky places throughout the southeastern United States

Georgia Rush

Juncus georgianus Coville
Rush Family

Description: A clumped grasslike perennial, the Georgia Rush has flat leaves that are only half as long as the terminal inflorescences. Although its perfect flowers are seldom seen in bloom, numerous plants will flower in unison on certain mornings, shed their pollen, and close by noon. Days of synchronous flowering are separated by intervals of several days before the plant flowers again. The capsular fruits have numerous seeds and are surrounded by six small greenish perianth parts.

Occurrence: Abundant in the shallow depressions on most granite outcrops where competition from other species is negligible.

Distribution: Endemic to granite outcrops from North Carolina to Alabama

Chinese Privet

Ligustrum sinense Lour.
Olive Family

Description: A naturalized woody species, the Chinese Privet often becomes an obnoxious weed that forms almost impenetrable thickets in disturbed habitats such as floodplains. Except when winters are severe, the small entire opposite leaves with short petioles remain green. The small branches are distinctively hairy and many terminate in branched inflorescences. The flowers are numerous, each with a four-lobed white corolla. The small bluish fruits are profuse in number and are readily consumed by birds who aid greatly in this plant's distribution.

Occurrence: Ubiquitous in the deep soils of rock outcrops, especially where the rock adjoins fields or second-growth woodlands.

Distribution: Widespread throughout the eastern United States

False Pimpernel

Lindernia monticola Muhl. ex Nutt.
Figwort Family

Description: A perennial, the False Pimpernel's leaves are mostly basal. It has stem leaves, or bracts, that are opposite and very small. The white to violet flowers are solitary and distinctly two-lipped, the upper lip being two-lobed, and the bottom three-lobed. The flowers are borne from the leaf axils from spring to fall if sufficient moisture is present.

Occurrence: At the margins of some depression pits that hold water for extended periods.

Distribution: On granite outcrops from North Carolina to Alabama, but rare elsewhere

Sundrops

Oenothera fruiticosa L.
Evening Primrose Family

Description: A perennial very common to granite outcrops, this species has showy yellow flowers and erect stems with oppressed hairs bearing alternate leaves that vary in size and shape. The flowers are solitary with four reflexed calyx lobes, four petals about 1 inch long, and four stigmas. The ovary of the pistil, where the plant's capsular fruit develops, is a great distance below the other flower parts.

In Sundrops, flowers open in sunshine, unlike many members of this family whose flowers open only in the evening, hence the common family name. A rare relative on the granite outcrops is *O. linifolia* Nuttall, a slender annual with linear leaves and small yellow flowers.

Occurrence: In the relatively deep soils of depression pits and border areas.

Distribution: Southeastern United States

Virginia Creeper

Parthenocissus quinquefolia (L.) Planchon
Grape Family

Description: A woody vine with alternate palmately compound leaves, the Virginia Creeper's usual five leaflets are coarsely serrate and grow up to 6 inches in length. Inflorescences with many inconspicuous flowers develop in late spring followed in fall by the maturation of black or dark blue fruits. This species trails along the ground or climbs high into trees by means of modified tendrils in the form of adhesive discs.

Occurrence: In deep soils on and around the granite outcrops.

Distribution: Southeastern United States

Spotted Phacelia

Phacelia maculata Wood
Waterleaf Family

Description: Phacelia plants are spring flowering annuals with alternate compound leaves. Radially symmetrical flowers with five petal lobes are borne on a terminal coiled inflorescence that unfurls like a fern's fiddlehead.

A related species, *Phacelia dubia* (L.) Small, Phacelia, is associated with granite outcrops in open border areas between rock and adjacent field or pine forest. In distinguishing between the two species, the blue to white flowers of *P. dubia* are the smaller of the two, and the darker lavender flowers of *P. maculata* have ten dark petal spots. In addition, the sepals of the latter have prominent bristles at their margins.

Occurrence: In the partial shade of glade areas between rock and deciduous forest.

Distribution: Endemic to granite outcrops of the Piedmont Region

Black Cherry

Prunus serotina Ehrh.
Rose Family

Description: The Black Cherry is a native tree that bears long inflorescences of numerous small white flowers that bloom before the leaves are fully developed. The small cherry fruits, coveted by birds, change in color from red to black as they ripen. Its alternate simple leaves are finely serrate along the margin, and its thin glossy twigs, which have horizontal lenticels, give off the stringent odor of bitter almonds when bruised.

Occurrence: One of the earliest trees to colonize depressions in the rock where the soil is deep enough to support woody perennials.

Distribution: Widespread in eastern North America

Hog Plum

Prunus umbellata Ell.
Rose Family

Description: A scraggly shrub or small tree in appearance, the Hog Plum has white flowers that are borne singly or in clusters and appear before the leaves in early spring. Each flower, measuring $^1/_2$ inch in width, has five petals, many stamens, and a single pistil. The alternate elliptic leaves have crenate to serrate margins and grow up to 3 inches in length. Dark purple to black fruits mature in late summer.

A related thicket-forming shrub, *P. angustifolia* Marshall, Chickasaw Plum, may also be found in association with granite outcrops. Its distinguishing features include glands at the tip of leaf serrations and large red or yellow fruits.

Occurrence: Grows in colonies in rock crevices or border areas.

Distribution: Southeastern United States

Wafer Ash

Ptelea trifoliata L.
Rue Family

Description: The Wafer Ash is a shrub or short-lived tree having alternate compound leaves consisting of three ovate leaflets. Clusters of small greenish-white flowers appear in late spring. They are foul-smelling and pollinated by carrion flies. By late summer, large round two-seeded fruits are evident in drooping clusters. The thin flattened fruits resemble a wafer, hence the common name.

Occurrence: Infrequent in occurrence, but found in sites where hardwoods border an outcrop.

Distribution: Eastern North America

Georgia Oak

Quercus georgiana M.A. Curtis
Beech Family

Description: A small scrubby tree, the Georgia Oak has green shiny leaves, 3 to 4 inches in length, with three to five bristle-tipped lobes. The base of the leaf blade, where it joins the leaf stalk, resembles an inverted triangle. The acorns of the Georgia Oak are small and have a cup that covers about one-fourth of the nut.

Other oak species, widespread and common in surrounding forests, sporadically appear in deep soils of granite outcrops. These include the following: *Q. nigra* L., Water Oak, whose small leaves are broadest near the top; *Q. falcata* Michx., Southern Red Oak, whose large bristle-lobed leaves are light-colored and hairy underneath; *Q. margaretta* Ashe, Scrubby Post Oak, which has small lobed leaves with hairs underneath but without bristle tips; *Q. marilandica* Muenchh., Blackjack Oak, whose large leathery leaves, shaped like inverted triangles, have shallow bristle-tipped lobes; and *Q. prinus* L., Rock Chestnut Oak, with many shallow rounded lobes that lack bristle tips.

Occurrence: Endemic to granite outcrops.

Distribution: Georgia, Alabama, and South Carolina

Sunnybell

Schoenolirion croceum (Michx.) A. Gray
Lily Family

Description: An attractive bulbous perennial, the Sunnybell has an 8 to 24 inch leafless stem that arises from a basal rosette of thin linear leaves and terminates in an inflorescence of many flowers. Each bright yellow flower is borne on a stalk accompanied by a small thin, dry leaf where it diverges from the stem. The flowers have six perianth parts united at their bases and six stamens.

Occurrence: In moist drainage areas and semiaquatic sites on most granite outcrops.

Distribution: Restricted to Georgia, Florida, and a few sites in the Carolinas

Cottony Groundsel

Senecio tomentosus Michx.
Aster Family

Description: A perennial, the Cottony Groundsel has yellow ray and disc flowers borne on a stout branched stem that grows between 1 and 2 feet in height. The underside of the unlobed basal leaves and the lower part of the stem are covered in soft dense hairs that give the appearance of white felt.

Occurrence: In large colonies in the moderately deep soils of depression pits.

Distribution: Found on all principal outcrops of the Piedmont Region but otherwise confined to the Coastal Plain Region of the southeastern United States

Catbrier

Smilax bona-nox L.
Lily Family

Description: A slender perennial vine that climbs by means of tendrils, the Catbrier has thorny stems that bear semievergreen alternate cordate leaves with parallel veins. Its leaves are variable in shape and have a thickened margin that is sometimes prickly. Both male and female plants bear small flowers in an umbellate inflorescence and produce an umbellate cluster of bluish-black spherical fruits in the fall.

A related species, *Smilax glauca* Walter, has leaves that are whitened on the underside and is also common to granite outcrops.

Occurrence: In deep soils of depression pits on granite outcrops.

Distribution: Eastern United States

Southern Slender Ladies' Tresses

Spiranthes gracilis (Bigel) Beck
Orchid Family

Description: The Southern Slender Ladies' Tresses is a perennial orchid whose leaves are basal and usually absent at flowering time. A slender stem, 1 to 2 feet tall, bears small white flowers in a single rank that is usually spiralled in appearance. It is the only member of the Orchid Family to inhabit granite outcrops.

Occurrence: In dry, open areas bordering the rock outcrop.

Distribution: Eastern United States

Hairy Spiderwort

Tradescantia hirsuticaulis Small
Spiderwort Family

Description: A densely hairy perennial herb that may grow up to 16 inches in height, the Hairy Spiderwort produces colorful flowers that range in color from purple to rose-red, and occasionally white. The veins of its linear leaves run parallel to one another. The flowers, conspicuous and radially symmetrical, have three petals and six hairy stamens and open for one day only. The Hairy Spiderwort may continue to produce flowers throughout the spring season.

A relative, *T. ohiensis* Raf., may also be present on the granite outcrops. It can be distinguished from *T. hirsuticaulis* by its larger size and lack of hairs.

Occurrence: On granite outcrops where plants can gain a foothold in cracks or ledges.

Distribution: In the Piedmont Region from North Carolina to Alabama

Sparkleberry

Vaccinium arboreum Marsh
Heath Family

Description: A slowly growing shrub or small tree with a profusion of thin crooked branches, the Sparkleberry has small simple alternate leaves that are shiny above and leathery in texture. Its leaves are either evergreen or turn a colorful red during mild winters. Small white bell-shaped flowers have ten stamens and are borne on leafy inflorescences. Its nonpalatable black berries mature in the fall.

Occurrence: A common member of all granite outcrops, it grows as a "pioneer" shrub where soil can sustain it.

Distribution: Throughout the southeastern United States, in a great variety of habitats

Muscadine

Vitis rotundifolia Michx.
Grape Family

Description: The Muscadine is a woody vine that climbs by means of its tendrils and is sometimes found trailing across the surfaces of outcrop rocks. Its simple alternate leaves are round in shape, spanning up to 3 inches in length and width, and are strongly toothed. The inconspicuous flowers that appear in late spring produce sweet edible purple fruits in the fall. One of the many varieties of Muscadine that is cultivated is the Scuppernong, known for its golden-green fruit.

Occurrence: In deep soils in and around the granite outcrops.

Distribution: Southeastern United States

Bear Grass

Yucca filamentosa L.
Lily Family

Description: A prominent and handsome plant, the Bear Grass can reach anywhere from 4 to 6 feet in height. Its unbranched stem arises from a basal rosette of long rigid evergreen leaves whose margins often appear frayed and showing long threads. The stem terminates with a large branched, many-flowered inflorescence. Large white pendulous flowers have six petal-like parts, six stamens, and a pistil with a three-parted stigma.

Bear Grass flowers are pollinated by the tiny white pronuba moth. The moth collects a mass of pollen, pushes it into the stigma, and oviposits eggs in the ovary that becomes a many-seeded fruit. This procedure is repeated many times; as the fruits mature, moth larvae eat some of the developing seed before finally exiting from the fruit through a bored hole.

Occurrence: In hot and dry habitats.

Distribution: Southeastern United States

Summer Flowering Plants

False Aloe

Agave virginica L.
Amaryllis Family

Description: The False Aloe is a summer flowering perennial adapted to desertlike conditions. It is distinguished by a tall stem, up to 6 feet in height, that arises from a basal rosette of leaves, and an unbranched terminal inflorescence of large (but not showy) flowers. Its basal leaves may have purple spots. The greenish to purplish six-lobed corolla is tubular for more than one-half of its length and six large yellow stamens extend beyond the lobes.

The False Aloe may be confused with the Yucca but the latter has a much larger branched inflorescence and pendulous flowers of six petal-like parts united only at the base. Some taxonomists place it in a different genus, *Polianthes virginica* (L.) Shinners.

Occurrence: Dry and sandy habitats bordering outcrops.

Distribution: Eastern North America: the only species of its genus to grow north of Mexico and Texas

Blackberry Lily

Belamcandra chinensis (L.) D.C.
Iris Family

Description: The Blackberry Lily displays a strikingly handsome flower that is borne on a 3-foot stem arising from clasping irislike leaves. The prominent sepals and petals are yellow-orange and mottled with purplish spots. At maturity, the dry capsular fruit opens to reveal a cluster of black seeds that resembles a blackberry, hence the common name. This perennial was naturalized from Asia.

Occurrence: The Blackberry Lily grows well in hot and dry locations. Rogers McVaugh notes that it is one of the few nonnative plant species that grows well on the granite outcrops.

Distribution: Widespread in the United States

Tufted Sedge

Bulbostylis capillaris (L.) Clarke
Sedge Family

Description: A diminutive grasslike annual, the Tufted Sedge has slender stems that are round in cross-section and coarsely ribbed. The basal leaves are slender, almost hairlike, and its flowers, which are lacking in petals, are borne in solitary several-flowered spikelets. The small nutlike single-seeded fruits of the Tufted Sedge are three-angled and bear a small bulblike protuberance.

Occurrence: In outcrop depression pits with very shallow soil.

Distribution: Throughout much of the United States and in other countries

Beauty-Berry

Callicarpa americana L.
Olive Family

Description: The Beauty-Berry is a shrub with hairy twigs that can grow up to 8 feet in height. Its ovate opposite leaves have a crenate to serrate margin and a hairy underside. In early summer, clusters of small pale bluish to reddish, radially symmetrical, four-lobed tubular flowers are clustered in the leaf axils. These are followed in late summer and fall by clusters of green fruits that turn a striking shade of magenta.

Occurrence: In woodlands bordering rock outcrops.

Distribution: Southeastern United States

Dayflower

Commelina erecta L.
Spiderwort Family

Description: The brilliance of this species in flower must be experienced early in the day as its large blue flowers open for only a single morning before wilting. A perennial with thick fibrous roots, its 6 to 24 inch tall stems bear alternate lanceolate to linear leaves with parallel veins and basal sheaths.

Sometimes the dayflower is confused with species of *Tradescantia*, Spiderwort, which also bear ephemeral morning flowers but whose three flower petals are equal in size and shape. Dayflower has bilaterally symmetrical flowers with two large rounded blue petals and one small whitish petal.

Occurrence: In deep soils adjacent to outcrops that have the capacity to retain moisture for long periods of time.

Distribution: Eastern United States

Coreopsis

Coreopsis grandiflora Hogg
Aster Family

Description: Perhaps the most common and colorful summer flowering perennial species of the granite outcrops, the Coreopsis' yellow ray flowers are conspicuously notched. The involucre of green bracts that occur below each flowering head are in two series or groups. This plant reaches 3 feet in height and its glabrous stems bear opposite pinnately dissected leaves. Coreopsis has wings on its fruit that are entire and its leaf stalks are hairy.

A variety, *saxicola* (Alexander) E. B. Smith, has been observed on Stone Mountain and other granite outcrop locations. Unlike the Coreopsis, *saxicola*'s achene fruits are deeply incised and the plants are not hairy. Crosses made between the Stone Mountain *saxicola* plants and the typical *C. grandiflora* plants yield fertile offspring.

Occurrence: Infrequently in thin soils of outcrops.

Distribution: Southeastern United States

Rushfoil

Crotonopsis elliptica Willd.
Spurge Family

Description: This annual species is readily distinguished by the silvery star-shaped hairs and scattered brown to red spots on the underside of its leaves. The Rushfoil's small leaves are borne on delicate branched stems between 4 and 20 inches in height and are elliptical in shape. The flowers, which are borne on terminal inflorescences, are small, unisexual, and inconspicuous.

Occurrence: Ubiquitous on granite outcrops, the Rushfoil grows in dry, sandy soils where competition from other plants is weak.

Distribution: Eastern United States

Granite Sedge

Cyperus granitophilus McVaugh
Sedge Family

Description: An abundant annual, the Granite Sedge is one of the few nonsucculent species that can withstand desiccation for long periods of time. Its tufts of triangular stems bear basal leaves keeled on the underside and clusters of flattened spikelets consisting of many two-ranked scales. The petalless flowers are borne in the axils of spikelet scales that turn from green to brown when the nutlet fruits are mature.

Occurrence: Endemic to granite outcrops, the Granite Sedge grows in depression pits with little soil and scant competition.

Distribution: North Carolina to Alabama

Poor Joe

Diodia teres Walter
Madder Family

Description: A hairy annual ranging between 10 and 30 inches in height, the Poor Joe has narrow opposite leaves that are less than 2 inches in length. Its flowers are small and white, and the petals are fused to form a tubular four-lobed corolla. Four stamens are attached to the tube of the corolla and the stigma is two-lobed.

A related species, *D. virginiana* L., Buttonweed, a common garden perennial weed, can be occasionally found on the outcrops. Its white flowers are about two times larger than Poor Joe's and its four corolla lobes are more abruptly spreading.

Occurrence: A native weedy species common on most granite outcrops, it also grows on dry, sandy soils and waste places.

Distribution: Southeastern United States

Pineweed

Hypericum gentianoides (L.) BSP.
St. John's Wort Family

Description: An abundant and characteristic species of outcrop communities, the Pineweed is a stiffly erect annual, ½ to 2 feet in height, with bushy branches and obscure scalelike leaves. Minute yellow, star-shaped flowers, scattered among the wing-angled branches, bloom in the morning and close up in the afternoon. Its annual associates, such as *Diamorpha* and *Arenaria*, flower in the spring whereas the Pineweed flowers from May to October and seedlings have been found throughout this period. Dead plants, which persist for some time, resemble diminutive trees.

Occurrence: Commonly found in depression pits with soil too shallow to support perennials except for lichens. It also grows in dry, sandy, shallow soils where most competitor species cannot survive.

Distribution: Eastern United States

St. Andrew's Cross

Hypericum hypericoides (L.) Crantz
St. John's Wort Family

Description: A low shrub that grows between 1 and 5 feet tall, St. Andrew's Cross has wing-angled branches and opposite linear to oblanceolate leaves that are narrowed at the base. Solitary flowers have four yellow petals whose shape somewhat resembles a St. Andrew's Cross. The species formerly was named *Ascyrum hypericoides* L.

Occurrence: In the herb-shrub community that lies between rock and adjacent forest.

Distribution: Widely distributed throughout the eastern United States

Shrubby St. John's-Wort

Hypericum prolificum L.
St. John's-Wort Family

Description: A woody shrub that may grow up to 6 feet in height, the Shrubby St. John's-Wort bears simple opposite, elliptic to linear leaves up to 2 inches long and ½ inch wide. A notch may be found at the base of the leaves and sepals. Its attractive flowers consist of five yellow petals up to ½ inch in length, numerous stamens, and a pistil.

Occurrence: Abundant where it occurs in rock crevices or in border areas between rock and woods. However, it is absent on many outcrops.

Distribution: Mountain and Piedmont regions of the southeastern United States

Pin-Weed

Lechea racemulosa Michx.
Rockrose Family

Description: A branching herbaceous perennial with stems up to 15 inches in height, the Pin-Weed displays opposite to nearly opposite, small narrow linear leaves less than 1 inch in length. Its branches arise from the leaf axils and bear numerous inconspicuous flowers. The fruits of the plant are small and pear-shaped and resemble the head of a pin, hence its common name.

Occurrence: In shallow, dry soil associated with granite outcrop communities.

Distribution: Southeastern United States

Flax

Linum virginianum L.
Flax Family

Description: A native perennial species related to the European flax of commerce, the Flax has slender stems that grow up to 3 feet in height and bear simple thin narrow leaves, mostly alternate in arrangement, and up to 1 inch in length. Its flowers are radially symmetrical with five yellow petals, five stamens, and a pistil with five stigmas.

Occurrence: In depression pits and border areas that have sufficient soil for herbs but not for woody plants.

Distribution: Eastern United States

Prickly Pear

Opuntia compressa (Salisb.) Macbr.
Cactus Family

Description: A native cactus species in the Piedmont Region, the Prickly Pear has a fleshy stem that is branched and prostrate. The jointed stem segments are sometimes armed with long spines and always armed with clusters of tiny bristles that readily and painfully penetrate the skin when touched. Its large flowers display many yellow petals and numerous stamens. The fruits of the Prickly Pear are purple and are edible after the outer covering is removed. Another species, *O. drummondii* Grah., common in the Coastal Plain, has been reported from Heggie's Rock. It has small stem sections with large spines.

Occurrence: Rocky or sandy habitats.

Distribution: Eastern United States

Rock Outcrop Milkwort

Polygala curtissii Gray
Milkwort Family

Description: A branched annual reaching between 1 and 2 feet in height, the Rock Outcrop Milkwort has alternate linear leaves and clusters of small bilaterally symmetrical flowers that terminate its branches. The flowers are rose-purple in color with yellow-tipped petals. Flowers bloom throughout summer depending upon adequate moisture.

Occurrence: In depressions in the rock where soil is shallow.

Distribution: Southeastern United States, especially in the Piedmont and Mountain regions

Small's Portulaca

Portulaca smallii P. Wilson
Purslane Family

Description: A succulent annual, Small's Portulaca is much branched with a spreading habit and narrow alternate fleshy leaves. Its small pink flowers, borne in terminal inflorescences, are almost hidden by a surrounding cluster of small leaves and long hairs. The flowers have two sepals, five petals, and eight to twelve stamens.

Another species in the same genus, *P. coronata* Small, occurs infrequently on the granite outcrops but is not endemic to them. It differs from Small's Portulaca in that its leaves are larger and spatulate-shaped and its yellow flowers are not closely surrounded by leaves and hairs.

Occurrence: Endemic to the granite outcrops, it is often the only living plant found growing in shallow soil during the hot dry summer months.

Distribution: North Carolina to Alabama

Meadow-Beauty

Rhexia mariana L.
Melostome Family

Description: A perennial with conspicuously hairy stems, the Meadow-Beauty has opposite leaves that are three-nerved and serrate. The purple to white flowers have four petals and eight prominent yellow curved stamens. The fruit is borne within an urn-shaped structure that is covered with glandular hairs.

Rhexia virginica L. occurs occasionally in outcrop communities. It may be distinguished from *R. mariana* by its conspicuously winged stem.

Occurrence: Inhabits relatively deep soils at the edge of the rock outcrop.

Distribution: Widespread in the eastern United States from Massachusetts to Florida

Winged Sumac

Rhus copallina L.
Cashew Family

Description: A short-lived shrub or small tree with alternate pinnately compound leaves, the Winged Sumac has eleven to twenty-three shiny leaflets with smooth margins. The rachis of the leaf has wings of leaflike tissue between the leaflets. *Rhus glabra* L., Smooth Sumac, also occurs on the outcrops. Its leaflets are more serrate than the Winged Sumac and the rachis is not winged. Both species have small flowers and conspicuous clusters of reddish fruits in a terminal inflorescence.

Another species, *Rhus aromatica* Ait., Fragrant Sumac, is sometimes found as a component of shady hardwood slopes adjacent to rock outcrops. When bruised, the plant emits a pleasant odor. The leaves of this small shrub have only three leaflets and often are mistaken for Poison Ivy, another species in the genus *Rhus*.

Occurrence: Commonly in thickets adjacent to outcrops.

Distribution: Eastern United States

Beak Rush

Rhynchospora grayi Kunth
Sedge Family

Description: The Beak Rush is one of several species of summer flowering perennials with leaves and stems arising from the base of the plant. The stems and branches bear loose clusters of small brownish spikelets with inconspicuous flowers. Several other species of *Rhynchospora* have been found on granite outcrops.

Occurrence: In shallow depression pits on the granite outcrops.

Distribution: Southeastern United States

Georgia Savory

Satureja georgiana (Harper) Ahles
Mint Family

Description: A low, sprawling shrub with pubescent twigs and aromatic opposite leaves, the Georgia Savory produces two-lipped mintlike flowers that range from white to rose in color. The corolla of fused petals has a two-lobed upper lip and three-lobed lower lip. Its four stamens and two-cleft stigma protrude beyond the corolla.

Occurrence: In the partial shade of hardwoods in the border area of outcrops.

Distribution: Southeastern United States

Rock Pink

Talinum teretifolium Pursh.
Portulaca Family

Description: The Rock Pink is a perennial with succulent terete alternate leaves that reach about 1.5 inches in length. Its pink flowers, borne in open branched inflorescences, open late in the afternoon and are ephemeral. These flowers consist of two green sepals, five small rounded petals, fewer than twenty stamens, and a short style whose stigma is level with the pollen-bearing anthers of the stamens. If cross-pollination does not occur, flowers self-pollinate.

Occurrence: In shallow soils adjacent to rock exposures on most granite outcrops in Georgia.

Distribution: From serpentine rock outcrops in eastern Pennsylvania to granite and sandstone outcrops in Georgia and Alabama

Menges' Rock Pink

Talinum mengesii Wolf
Portulaca Family

Description: This species is similar to the more common *T. teretifolium* Pursh., or Rock Pink, in its vegetative aspect and open inflorescence. However, its ephemeral pink flowers open early in the afternoon, about two hours before those of the Rock Pink. The Menges' Rock Pink produces flowers that have two green sepals, long acute petals, more than forty stamens, and a long style whose stigma is above the level of the stamens. If pollination does not occur, flowers do not self-pollinate.

Occurrence: In shallow soils adjacent to both granite and sandstone outcrops.

Distribution: Sandstone outcrops of northern Alabama and Tennessee and on some granite outcrops in Georgia and Alabama

Yellow-Eyed Grass

Xyris jupicai Richard
Yellow-Eyed Grass Family

Description: A perennial herb with flat and narrow basal leaves that resemble those of the Iris, the Yellow-Eyed Grass has a leafless stem, 1 to 2 feet in height, with two prominent ribs. It terminates in a round, compact inflorescence that has closely packed hard scales. During the summer, bright yellow flowers emerge from behind the scales. These flowers are ephemeral and consist of three sepals, three yellow petals, three stamens, three stigmas, and a many-seeded pistil. The fruits and seeds of the Yellow-Eyed Grass develop behind the persistent scales.

Occurrence: At the edge of pools and seeps with shallow soil. Several species of *Xyris* have been reported growing in rare wet depressions on granite outcrops.

Distribution: Southeastern United States

Late Summer and Fall Flowering Plants

Slender Gerardia

Agalinus tenuifolia (Vahl) Raf.
Figwort Family

Description: The Slender Gerardia is a many-branched annual with slim green to reddish stems that bear narrow opposite leaves. Its showy purple flowers are slightly two-lipped with the upper lip obscuring the four stamens. The throat of the light purple corolla has two yellow pollinator guide lines and is not hairy like other related species.

Occurrence: The border area between rock and woodland, flowering in both late summer and fall.

Distribution: Eastern United States

Broom Sedge

Andropogon virginicus L.
Grass Family

Description: The Broom Sedge is a tufted perennial grass that grows between 2 and 3 feet tall. A native grass common in recently abandoned farmland, it flowers in the fall. The spikelets, bearing inconspicuous flowers, are surrounded by silky hairs. Throughout the winter and spring, the dried plants can be identified by their yellowish-brown color.

A similar species, *A. scoparius* Michx., Little Bluestem, occurs infrequently on the outcrops. Its visibly jointed stem nodes are purplish in color.

Occurrence: It is one of the first perennials to become established in the deep soil of rock depressions.

Distribution: Southeastern United States

Alexander's Rock Aster

Aster avitus Alexander
Aster Family

Description: A composite-flowered perennial with alternate simple narrow leaves, the Alexander's Rock Aster produces a conspicuous daisylike head of many small disc flowers surrounded by large strap-shaped ray flowers. The latter are less than thirteen in number and lavender in color. A distinguishing feature beneath the flowers is the collection of overlapping bracts that have green tips that tend to flair outward.

Occurrence: In patches in the grassy margins of outcrops.

Distribution: Occasionally found in outcrop communities in Georgia and South Carolina

Confederate Daisy

Helianthus porteri (A. Gray) Pruski
Aster Family

Description: The brilliance and profusion of these blazing yellow flowers has inspired the Yellow Daisy Festival held each year at Stone Mountain Memorial Park in DeKalb County, Georgia. The Confederate Daisy is a late summer annual that may reach 2 to 3 feet in height and bears opposite hairy linear leaves. The composite head of its flowers has an involucre of overlapping bracts and showy yellow ray flowers. The seeds of this species germinate in spring and plants require the entire summer to reach reproductive maturity. The Confederate Daisy formerly was known as *Viguiera porteri* (A. Gray) Blake.

Occurrence: In soils sufficiently deep for some water retention but shallow enough to prevent perennial herbs and shrubs from becoming established.

Distribution: Endemic to granite outcrops in Georgia and a few locations in Alabama and introduced to outcrops in North Carolina, including Rocky Face, in 1959

Blazing Star

Liatris microcephala (Small) K. Schumann
Aster Family

Description: A most handsome perennial, the Blazing Star has erect stems that may grow between 1 and 3 feet in height, and displays alternate narrow entire leaves below and a cylindrical mass of purple flowers above. Its flowers are borne in three- to five-flowered involucral heads that are attached to the stem by a short stalk. It is distinct from two close relatives, *L. graminifolia* (Walt.) Willd. and *L. spicata* (L.) Willd., by its small number of flowers per head. The Blazing Star is a flowering companion to the Confederate Daisy.

Occurrence: In border areas between flat-rock surfaces and adjacent forest.

Distribution: Ranges from North Carolina and Kentucky to

Flatrock Panic Grass

Panicum lithophilum Swallen
Grass Family

Description: Flatrock Panic Grass is a small tufted annual grass that can grow up to 10 inches in height. Usually it has several purplish branches that terminate in a delicate highly branched inflorescence. The erect purplish leaf blades grow up to 3 inches in length and 1/4 inch in width. Several other species of *Panicum* are common on the outcrops, including the larger perennial *P. meridionale* Ashe, which flowers earlier.

Occurrence: Often the only plant to be found in shallow soils of depression pits in late summer.

Distribution: A rare species confined to the granite outcrops of Georgia

Cone Bearing Plants

Eastern Red Cedar

Juniperus virginiana L.
Cypress Family

Description: The Eastern Red Cedar is a medium-sized evergreen tree very common on most granite outcrops. Its hard aromatic wood is used for fence posts and cedar chests, and its bluish berrylike seed cones, produced in the fall, provide food for wildlife. The bark of the Eastern Red Cedar is brown and shredded in appearance. Its small scalelike leaves cover the twigs that appear four-sided in cross-section.

Occurrence: Varying in size and shape, these trees grow in crevices in the rock and in border areas between rock and woodland.

Distribution: Southeastern United States

Loblolly Pine

Pinus taeda L.
Pine Family

Description: Abundant in association with the outcrops, the Loblolly Pine has needlelike leaves that are borne three per cluster and can reach 7 or more inches in length. Deciduous male cones shed pollen in early spring. Female cones, which persist for several years, reach 8 inches in length and have stout prickles.

Two other pine species occur on outcrops infrequently: *P. echinata* Miller, Shortleaf Pine, and *P. virginiana* Miller, Scrub Pine. The former bears needles that are between 4 and 5 inches in length and two per cluster while the latter has short twisted needles, 3 to 4 inches in length, and two per cluster.

Occurrence: Saplings occur in the deeper soil of depression pits but persist to a large size only where roots can penetrate cracks and crevices in the rock.

Distribution: Common in the southeastern United States and ranging from Texas to New Jersey

Seedless Vascular Plants

Ebony Spleenwort

Asplenium platyneuron (L.) Oakes
Fern Family

Description: An evergreen perennial fern, the Ebony Spleenwort grows fronds up to 18 inches in height. The plants are commonly found growing in a colony. The fronds consist of a shiny black to brown stipe bearing alternate finely toothed leaf blades about 1 inch long. Fertile fronds bear two rows of spore sacs on the undersides of the leaf blades. The spore-bearing sacs are in an elongate cluster covered on one side by a transparent flap of tissue visible with a hand lens.

Occurrence: The Ebony Spleenwort can be found on rock outcrops in the partial shade of small trees.

Distribution: Eastern North America

Black-Spored Quillwort

Isoetes melanospora Engelm.
Quillwort Family

Description: Quillworts are spore-bearing perennials with round, narrow, hollow leaves that arise from a persistent underground stem and bear either small male spores or large female spores in sacs at their enlarged leaf bases. One of three quillworts associated with the granite outcrops, the Black-Spored Quillwort has long been considered a rare species endemic to granite outcrops. It may be a variant of the ubiquitous *I. piedmontana* Reed.

The amphibious *I. piedmontana,* the most common of the three Quillworts, grows in moist soil along rock margins where seasonal seepage paths occur, and also in deep muddy pools. Like *I. melanospora,* plants of this species are solitary with leaves arising from a two-lobed underground stem.

Another species, *I. tegetiformans* Rury, or Mat-forming Quillwort, identified in only two Georgia counties, is unique among quillworts with respect to elongate flattened stems with leaves produced only on one side. Its growth and vegetative reproduction results in a matlike tangle of plants within a depression pit.

Occurrence: Restricted to vernal pools that have a thin layer of soil and sufficient water-retention capacity to ensure an aquatic habitat for winter and early spring months. A common associate is the rare Pool Sprite.

Distribution: Alabama, Georgia, and South Carolina

Rock Spikemoss

Selaginella rupestris (L.) Spring
Spikemoss Family

Description: An evergreen spore-bearing, rock-inhabiting, vascular plant, the Rock Spikemoss tends to grow in mats. Its stems have a low creeping aspect, similar to the mosses with which it often grows, and numerous forking branches. The overlapping leaves are small, narrow, and gray-green with white terminal bristles and marginal hairs. Spore-bearing sacs at the bases of some leaves release the spores in the summer.

Occurrence: In the depression pits of granite outcrops and in sandstone locations.

Distribution: An Appalachian Mountain and northern species, occuring as far north as Canada and as far south as the Piedmont Region of Georgia

Large Woodsia

Woodsia obtusa (Sprengel) Torrey
Fern Family

Description: One of the few fern species common to the granite outcrops, the Large Woodsia produces clusters of pinnately divided, semievergreen fronds, 10 to 20 inches in length, that arise from creeping, perennial rhizomes. The fronds are scaly and brown in color. Spore-bearing sacs, apparent on the undersides of fertile fronds, occur in round clusters and are surrounded by semitransparent, ribbonlike protective tissue.

Occurrence: In the border areas between rock and woodland in partial shade, but can occasionally occur in full sun.

Distribution: Grows throughout the United States in association with dry ledges and rocky woodlands

Illustrated Glossary

LEAF CHARACTERISTICS

SIMPLE LEAVES

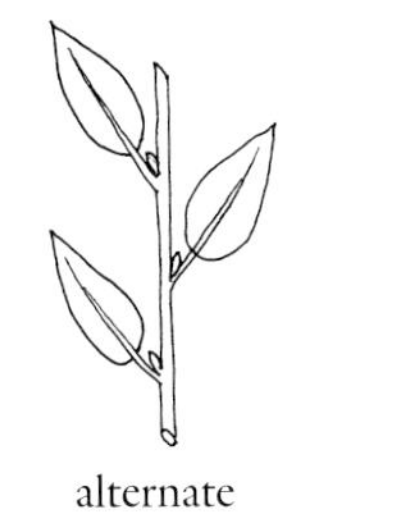

alternate

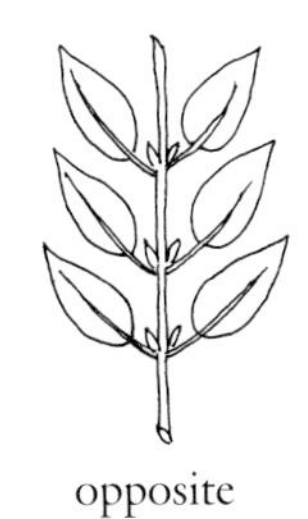

opposite

COMPOUND LEAVES

alternate

opposite

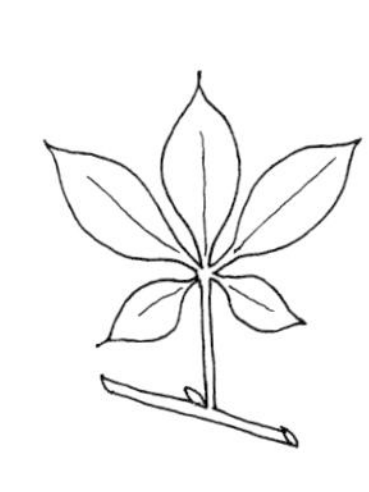

palmately compound

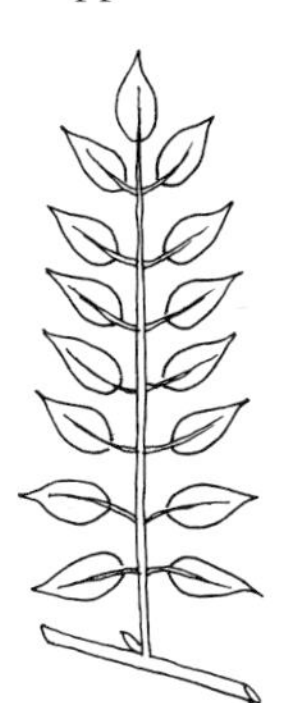

pinnately compound

LEAF SHAPES

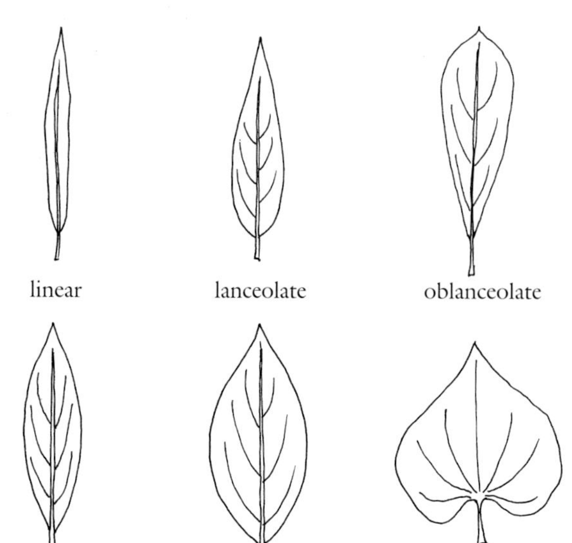

LEAF MARGINS

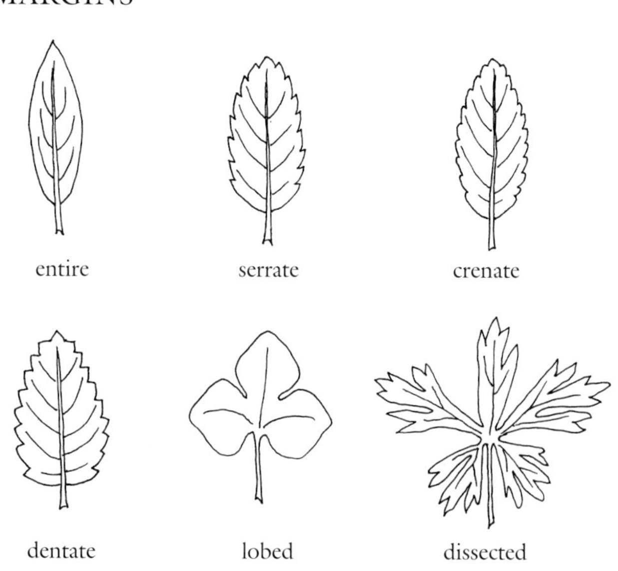

FLOWER CHARACTERISTICS

PARTS OF A FLOWER

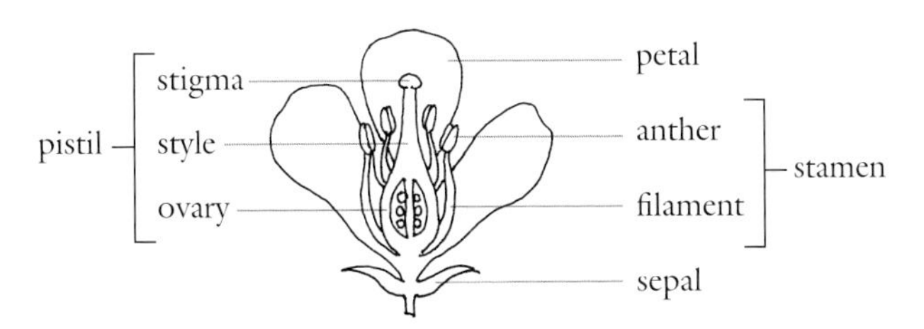

COMPOSITE FLOWER

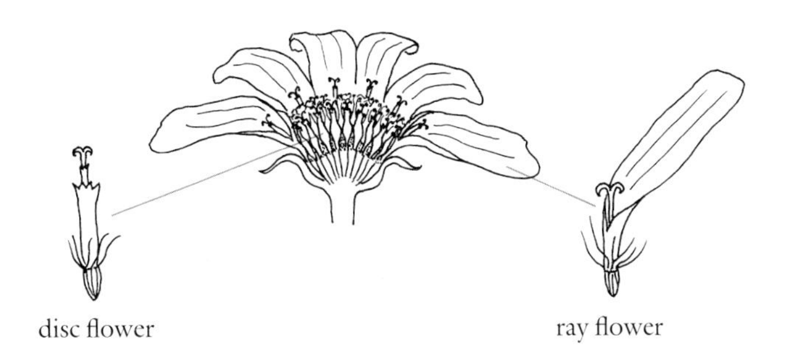

FLOWER SYMMETRY

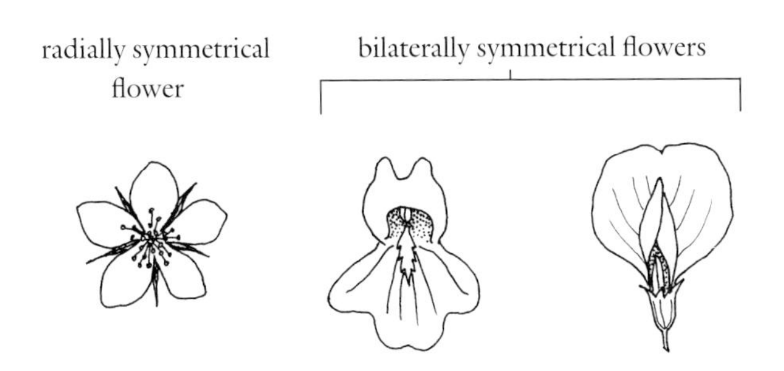

Glossary

Achene: A small seedlike dry fruit.

Annual: A plant that goes from seed to reproductive maturity to death in one year.

Awn: A slender appendage arising from a scale in the inflorescence of grasses.

Axillary: Borne from the upper angle between stem and leaf.

Biennial: A plant that lives for two years and flowers in the second year.

Bract: A small leaf associated with an inflorescence.

Calyx: The sepals collectively, either fused or separate.

Capsule: A dry many-seeded fruit.

Coastal Plain: The extensive geologically and ecologically variable region southeast of the Piedmont.

Composite: A headlike inflorescence of many small flowers that may all be similar or different, occurring in the form of disc and ray flowers.

Cone: A spore-bearing structure with variously arranged scales.

Corolla: The petals collectively, either fused or separate.

Crystalline rock: Rock that has been melted and crystallized upon cooling.

Deciduous: Absent or brown leaves that are nonfunctional in the winter.

Disc flower: The nonshowy central flowers of a daisylike composite flower.

Endemic: Restricted to a specific region or ecosystem.

Ephemeral: Short-lived, lasting but a day.

Evergreen: Leaves that remain green and functional in the winter.

Glandular hairs: Hairs with a round terminal gland visible with a hand lens.

Herbaceous: Plants that live only one year or die back each year to a perennial bulb, root, or corm.

Inflorescence: The flower-bearing part of a plant. It varies greatly in shape, size, structure, and number of flowers.

Involucre: A collection of bracts immediately below the flowers of a composite flower.

Lenticel: Corky light-colored lines or spots on young twigs.

Perennial: A plant that persists for more than two years.

Piedmont: A geologic province that stretches from New Jersey to Alabama.

Precambrian rock: Rock older than 600 million years.

Ray flower: The usually large and colorful peripheral flowers of a composite flower.

Rosette: A collection of leaves close to the ground.

Scale: A small leaflike structure.

Sheath: The portion of a leaf that surrounds the stem.

Spatulate: Spoonlike in shape.

Spikelet: A series of overlapping scales that enclose grass or sedge flowers.

Stipe: The central axis of a fern frond.

Succulent: A plant with thick fleshy leaves or stems.

Tendril: A slender modified stem or leaflet that coils around structures, enabling a plant to climb.

Tufted: Many stems arising from the base of a plant.

Umbel: An inflorescence where the flower-bearing branches originate at the same place.

Woody: Trees or shrubs with woody stems that remain alive in winter and produce new growth in the spring.

Selected References

Burbanck, M. P., and D. L. Phillips. "Evidence of Plant Succession on Granite Outcrops of the Georgia Piedmont." *American Midland Naturalist* 109 (1983):94–104.

Burbanck, M. P., and R. B. Platt. "Granite Outcrop Communities of the Piedmont Plateau in Georgia." *Ecology* 45 (1964):292–306.

Duncan, W. H., and M. B. Duncan. *Trees of the Southeastern United States*. Athens: University of Georgia Press, 1988.

Duncan, W. H., and M. B. Duncan. *Wildflowers of the Eastern United States*. Athens: University of Georgia Press, 1999.

Duncan, W. H., and L. E. Foote. *Wildflowers of the Southeastern United States*. Athens: University of Georgia Press, 1975.

Foote, L. E., and S. B. Jones. *Native Shrubs and Woody Vines of the Southeast*. Portland, Oregon: Timber Press, 1989.

McVaugh, R. "The Vegetation of the Granitic Flatrocks of the Southeastern United States." *Ecological Monographs* 13 (1943):120–61.

Radford, A. E., H. E. Ahles, and C. R. Bell. *Manual of the Vascular Flora of the Carolinas*. Chapel Hill: University of North Carolina Press, 1968.

U.S. Fish and Wildlife Service. "Recovery Plan for Three Granite Outcrop Plant Species." Jackson, Mississippi: U.S. Fish and Wildlife Service. 41 pp., 1993.

Wharton, C. H. *The Natural Environments of Georgia*. Atlanta: Georgia Department of Natural Resources, 1977.

Wyatt, R., and N. Fowler. "The Vascular Flora and Vegetation of the North Carolina Granite Outcrops." *Bulletin of the Torrey Botanical Club* 104 (1977):245–53.

Index to Common and Scientific Names